Connections to the Cosmos

George E. Hunter

Copyright © 2013 George E. Hunter

All rights reserved.

ISBN: 1490382291
ISBN-13: 978-1490382296

Picture 1. Astronomy can be a lifelong interest for people of all ages.

"Exploring the darkness is often lit by the flame of human imagination and curiosity". GH.2002

Picture 2. A small picture — yet of unimaginable vastness.

My thanks goes to Frann Leach who gladly slaved away at the typing, layout and some of the artwork for me, and to all the people who manage to read this book fully to the end.

Table of Contents

Introduction..7
Foreword..9
Billions...10
Gravity...10
Light..16
The Sun..19
Atoms..22
Stars..24
Space...32
Time..39
Galaxies...42
Quantum Mechanics...48
Planets...51
Comets, Meteors and Asteroids..69
The Future...73
Summary...75
Photo and Illustration Credits...77

Warning! — This book contains some scenes of a scientific nature — viewer discretion is advised!

Picture 3. If you're ready, your journey to the Cosmos starts here.

Introduction

Since the dawn of our species, humans have gazed skywards admiring the splendour of the cosmos, paying homage to its greatness and pondering on its deepest mysteries.

In the following pages, I wish to share some of the interesting and amazing facts about the universe and show how our expanding knowledge about it is helping to shape our lives for the future.

I believe, that we all have an innate curiosity about the galactic environment on our doorstep, but often find the sheer vastness of it all rather daunting. I trust the reader won't be discouraged by this, but instead find something to inform, entertain or astonish, regardless of your level of interest.

Take this journey through the cosmos at your own leisure. Explore the incredible power of a supernova, wander through the majestic Solar System, discover the strange properties of space and time, marvel at the beauty and nature of light, learn about the mind-blowing weirdness of the quantum world, or simply gaze at the nice pictures. The most important thing, is that I hope you find this book enjoyable!

Picture 4. Carl Sagan, award-winning astrophysicist, broadcaster and writer, who inspired my interest in astronomy.

Picture 5. Looking like a Disneyland castle, this interstellar cloud known as the Eagle nebula lies 7500 light years away in the Carina constellation. Clouds like these are where new stars and planets are being formed.

Foreword

As a small child, I would often gaze upwards at the night sky and was always spellbound by the awesome beauty and vastness of it all. What were these glistening specks of light that filled me with wonder? Why did they seem so far away? I would reach out my tiny hand in the futile hope of catching hold of these elusive beacons.

I wasn't aware back then that these celestial objects were distant suns that lay many light years away, or that our own Sun was also a star that provides the energy for all life here on Earth. I didn't know that our Solar System was originally formed from the remnants of dying stars that ejected their elements out into space or that all the atoms in my body were made deep within the cores of their stellar furnaces. Now I realize... Wow! I'm actually created from stuff that was produced in some far off twinkly lights!

If it has just come as a revelation that you are nothing but stardust, then next time you look up at the constellations, remember that you are part of the universe in a more intimate way than might be imagined, and that we are all inextricably linked through our **Connections to the Cosmos**.

Picture 6. The famous horsehead nebula in Orion. The red glow is ionised hydrogen gas caused by a nearby radiating star.

Billions

A Billion in this book is the international billion, written 1,000,000,000 or 1 followed by nine zeros which equals a thousand million.

A Million refers to 1,000,000 or a one with six zeros after it, or the equivalent of a thousand thousand.

Have you ever considered what large numbers like millions or billions really mean? Naturally it is extremely difficult to comprehend such figures, or even imagine them. If you started to count to a billion at regular one second intervals (without taking any rest or sleep) how long might it take you to complete that task?

The answer is not in hours or days but approximately 37 years or half a lifetime — 2,336, 2,337, yawn, 2,338, — zzzzz

Gravity

We go about our daily lives both accepting and using the force of gravity without too much thought about what it really is. We know that it causes objects to fall to the ground, and we use its force in our various recreational pursuits, like skiing downhill, going on a roller coaster or skydiving.

It was Sir Isaac Newton who speculated that the force acting on a falling apple was the same force that keeps the planets in orbit. He considered firing a cannonball at a specific speed. He reasoned that if you launch it fast enough and high enough, it would continue going all the way around the world.

The cannonball would be falling, but never actually landing, due to the curvature of the Earth. Newton concluded that the orbit of the Moon must behave in the same manner as the cannonball.

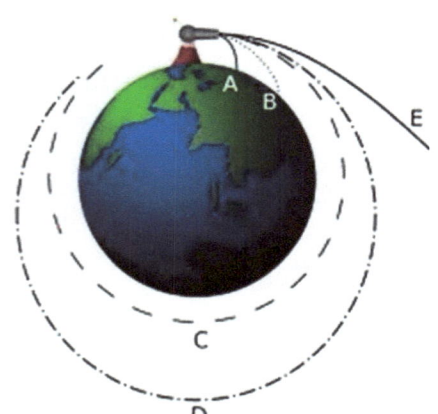

Picture 7. Given the right speed, the cannonball would take path C and be in orbit.

Imagine holding two pistols in your outstretched hands. You fire one and drop the other at the same time. Which will hit the ground first, the gun you released or the bullet from the barrel of the other?

Surprisingly, they land at exactly the same time! Regardless of the horizontal speed of the bullet, gravity acts equally on both objects. I'd never have worked all that out, even if an apple had reputedly fallen on my head!

Picture 8.

Although we have continually studied the effects of gravity, it is still one of the least understood forces in nature.

Newton also calculated that gravity is an attraction between two objects, and that the strength of gravity depends on the mass of the objects and diminishes with the square of the distance between them.

That being the case, I worked it out, that you would actually weigh less if you lived at the equator or on top of a mountain — although I'm not recommending this as a crash diet for slimmers!

So what might happen if there was no gravity at all? Well, apart from the fact that you would end up flying off into space at 1,000km an hour, as there would be nothing holding you down, there wouldn't have been any stars, planets or galaxies, and life wouldn't have evolved in the first place.

Gravity clumps things together, such as the stars, planets and galaxies, so without it Earth or the Sun wouldn't have formed.

Picture 9. Strange things can occur with gravity.

In a weightless environment, your heart becomes smaller and you actually grow taller. Also your muscles and bones can become weaker if not given regular exercise.

Picture 10.

Jack Lousma, who was one of the astronauts on Skylab, had just returned to Earth after spending several weeks there. Back home, he was putting on some aftershave when he let go of the bottle, and it ended up smashing on the floor.

It appears that he was expecting it to float, as it would normally have done in space. It seems it can also take some time for astronauts to re-adjust, after coming back down to Earth.

If there was no gravity at all, the Earth would literally blow itself apart. Throughout the Universe, gravity tries to squash anything with mass down to the smallest space possible. However, there are other forces which oppose this action (Newton's Third Law, that for every action there is an equal and opposite reaction).

Just as your body pushes back against the force of air pressure, the Earth's interior pressure pushes outwards against gravity. Without this balance, the Earth would either collapse in on itself or rip itself apart.

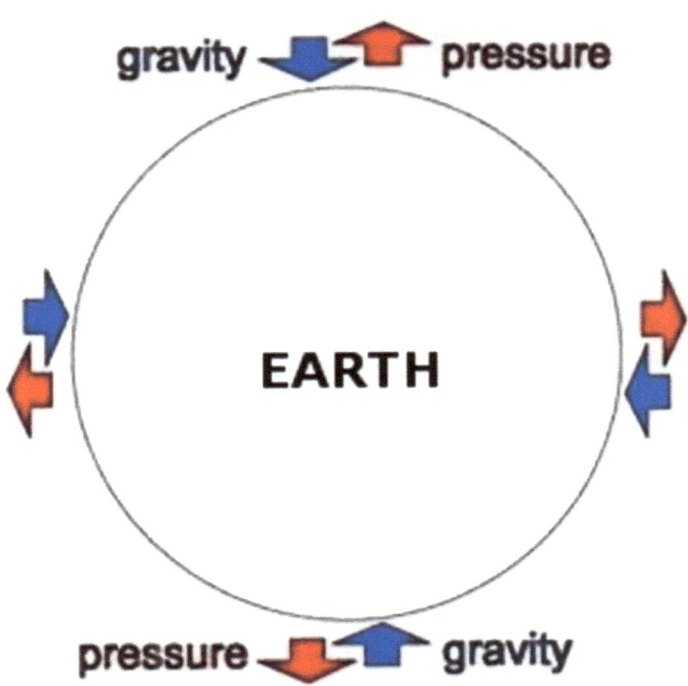

Picture 11. *Without the equilibrium between gravity and pressure, stars or planets would simply disintegrate.*

Newton's theory on gravity held sway for 150 years or more and helped to explain the orbits of the planets.

However, his equations did not really stand up in all circumstances, and so along came a young man to fiercely challenge these ideas and give a kick in the ass to the conventional thinking of the time.

That young man was none other than Albert Einstein. His brilliant (if not weird) idea was to suggest that gravity is not a force at all, but merely a distortion of space.

He visualised gravity as the result of the bending of space by various planets or any other objects with mass.

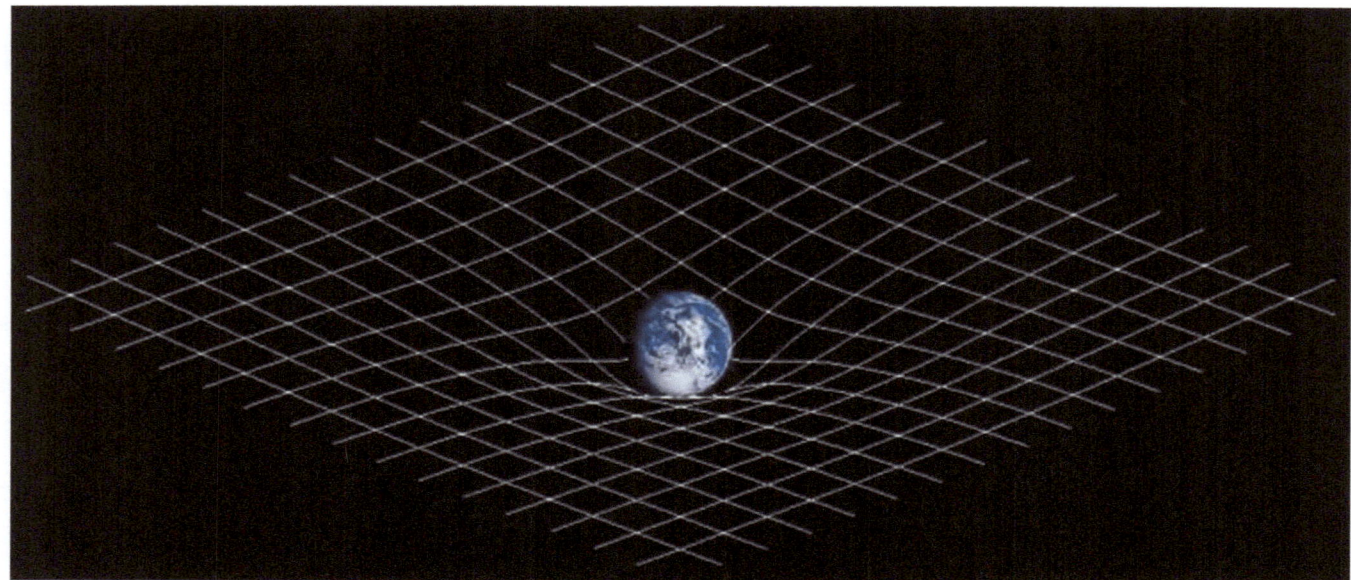
Picture 12. As Earth has mass, it can curve space, as shown above.

According to Einstein, it's this bending or curving of space that we experience or feel as gravity. It's not a force pulling you down but it's the curving of space that's really pushing you down. It's simply a question of visualising space as a structure or fabric that we are surrounded by.

If you feel that I'm losing you already. Take a look at the diagram and see if that helps. Whether or not you can get your head round it, all that matters is that his theory is our best explanation of gravity at present.

Even this ingenious concept breaks down though, when dealing with particles at the subatomic (very small) scale or in some of the most violent regions in the universe, such as the extreme gravity of black holes.

A little known fact about Sir Isaac Newton, was that his excellent powers of observation helped to unmask around 20 counterfeiters, during his occupation as Master of the Mint.

So how strong is gravity? Well, it's surprisingly rather weak. Using a small hand-held magnet, you can lift a set of car keys up off a table. This means that the magnet is strong enough to overcome the full force of gravity. Yet what's actually creating the gravitational force - is the whole Earth itself!

The mass of Earth is around six million billion billion kilograms, yet its gravity still isn't strong enough to overcome the force exerted by the magnet, whose mass is probably less than one kilogram. In other words, you can overcome the gravitational force created by the mass of the entire Earth with very little effort entailed.

Why gravity is so weak, is still one of the mysteries scientists are trying to resolve.

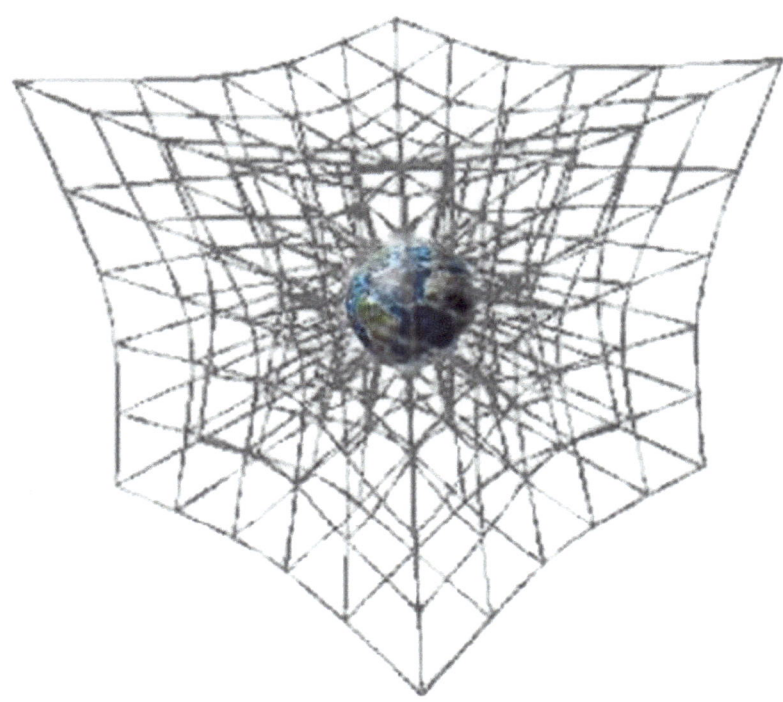

Picture 13. This squashing of the space around you is what's holding you down on Earth, not a force pulling you down.

All objects have a certain attraction to each other. Although it's hard to believe, right now you are exerting a gravitational force on the planet Neptune and it's four and a half billion kilometres away. In turn it's exerting a gravitational force on you. In fact, it's very pleasing to realise, that everyone on Earth is actually attracted to you, even if just gravitationally.

We can't see gravity but we do experience its effects. Everyday, the tides rise and fall due to the gravitational pull of the Moon.

Gravity also affects plant growth and even your body. From late teens onwards, your spine compresses making you smaller, your waist gets bigger as your internal organs basically drop downwards and your circulatory system gradually deteriorates.

Just as water can't flow uphill, neither can your blood circulate so easily, once you get older. Poor circulation caused various problems in astronauts too with their ears, limbs, eyes and even brains.

So where does that leave us? What exactly is gravity? We know that gravity definitely isn't responsible for people "falling in love", but the answer is, we don't precisely know. Until a new theory comes along that manages to explain it more fully and can unite both Newton and Einstein's theories into one Grand Unified Theory, we should just be glad that our feet are still held firmly on the ground.

Light

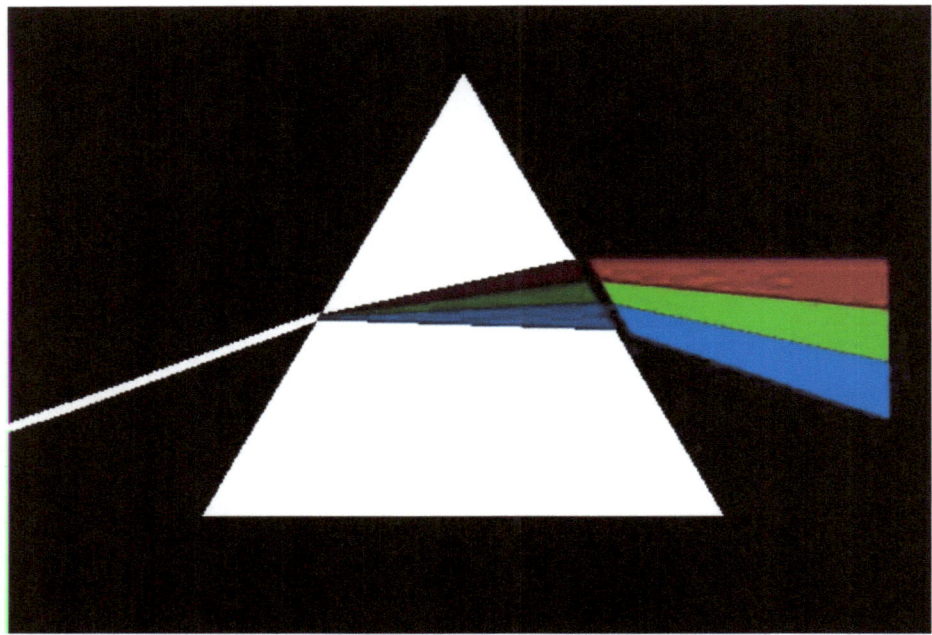

Picture 14. The colour white doesn't really exist.

When you view the moon or the stars in the night sky, you can't be certain they're still there. The reason is that the light from them takes time to reach your eyes.

If the moon was to suddenly explode, then you wouldn't be aware of it until about one second later. The nearest stars lie much further away though. If a similar fate occurred with Polaris (the northern pole star) it would be 430 years before you noticed it, and if it was a star in the Andromeda galaxy, you wouldn't know about it for around 2 million years!

In other words, the further away a star or galaxy is, the longer its light takes to reach us. In this way, we are able to look back in history to the earlier universe, which has contributed greatly to our overall understanding of it.

Furthermore, it has allowed us to view stars and galaxies of different ages, and from that, astronomers realised they go through various life cycles. This unique situation is comparable to being able to see a photo album of people from babies to adults and working out the connection.

You may know that light (photons) takes about eight minutes to reach us from the Sun. What might surprise you though, is that the journey photons take within the core of the Sun to reach its surface can last around 20,000 years or more.

As the photons embark upon their epic voyage, they bump into many other fast moving particles, causing them to take a very random path, just like the ball in a pinball machine, before finally exiting.

It was Albert Einstein who stated that light has a finite speed, and that it's not possible to exceed this limit. So far this has been proved to be correct.

So how fast does light actually travel?

It was in 1638 that the Italian, Galileo first attempted to measure the speed of light. He decided to use a lamp with a shutter mechanism attached to it. The idea was to have a friend stand some distance away using a similar device. When Galileo exposed the light from his lamp, his friend would open his shutter on receiving the signal.

Using this method, Galileo could measure the time delay to calculate the speed. After several attempts, however, he finally concluded that light travels "incredibly fast!"

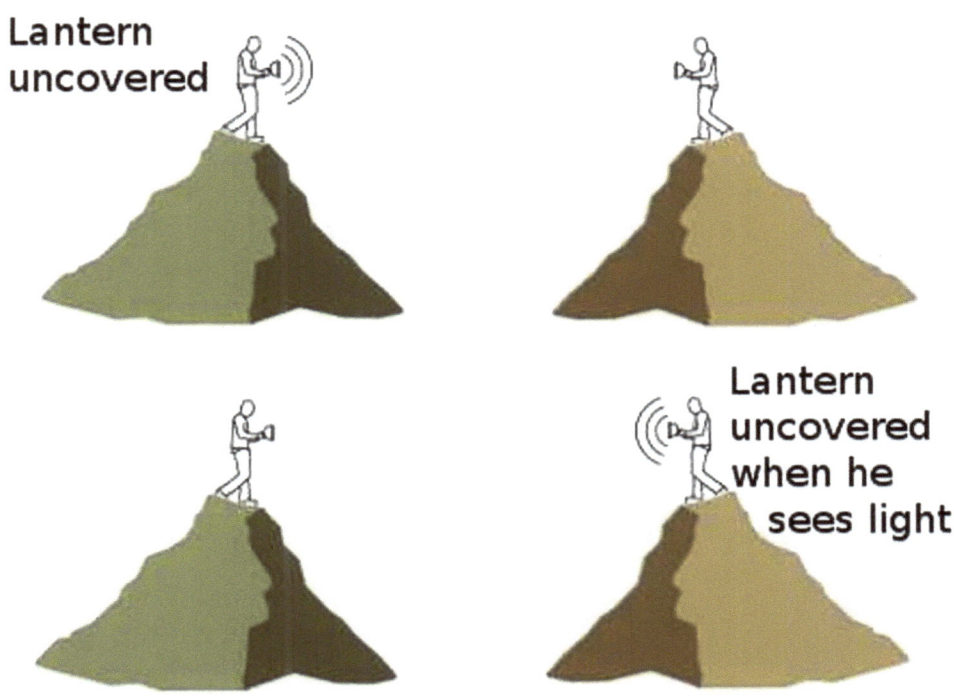

Picture 15. A representation of Galileo's speed of light experiment.

Due to the vastness of space, distances to the stars are not measured in metres or kilometres, but in light years — or the distance light travels in one year. Light travels at 300,000 kilometres per second, or 9.5 trillion (9,500,000,000,000) kilometres in one year, which is obviously too big a number to write down, so we use the light year instead.

To put that number into perspective, imagine you were standing at the equator and could travel at the speed of light. How far round the world do you think you might go in one second?

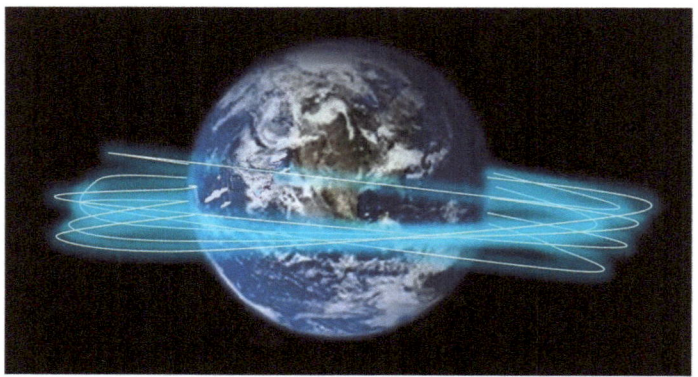

Picture 16.

If you have done the maths, the answer is a staggering seven and a half times!

That speed is only valid for light traveling in a vacuum though, and varies depending on the material, substance or object it happens to pass through. For example, it slows down in a liquid — which causes a straw in a drink to appear bent as the light rays are deflected.

If there wasn't a delay in the light reaching your eyes (from the absorption or reflection of different objects), then you would not be able to see the world in colour, and everything around you would simply be a blur.

Picture 17. About 150 years ago, there was no electricity, yet this picture shows the extent of our modern dependency on it.

Did you know, that our eyes are capable of distinguishing around 10,000,000 different shades of colour? No wonder some people have difficulty in choosing what to wear!

The Sun

Picture 18. Earth looks small, yet our Sun is only an average sized star.

The Sun is our nearest star, and it's so big you could fit over 100 Earth sized planets across its diameter, or put a million of them inside its sphere.

It's an immense ball of hydrogen gas which is gradually being converted into helium. It achieves this mainly by the intense heat and pressure within the Sun's core forcing the atoms together. Vast amounts of energy are released in this process. This is the same principle on which the nuclear bomb was modelled.

The Sun loses around 4 million tons of energy every second through the conversion process, yet it still has enough left over to keep it burning for another 5 billion years.

Although we are 150,000,000km away, we can still feel its awesome heat and can be illuminated by its penetrating light 100 metres below the ocean.

It's strange to think, that there's enough power released by the Sun each second to sustain the whole of the UK's energy demand for about the next 15 million years.

All that power, and yet some people are still struggling to pay their energy bills!

Picture 19. The dark spots on the Sun are caused by magnetic fields which make parts of the surface cooler or darker-looking.

Just as a driver relies on the engine of their car, then all life on this planet depends on the motor that keeps it running. The Sun is that provider and we sit at just the right distance away to take advantage of its benevolence. It's why Earth is often referred to as being in the "Goldilocks zone" — not too hot or cold. It's not always favourable to us though, as it can eject harmful radiation we call the solar wind, which can affect satellites, cause power cuts, change climate and even cause skin cancer.

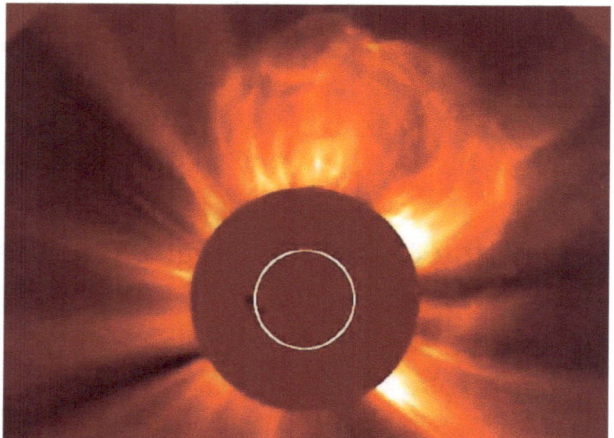
Picture 20. The solar wind (material) from the Sun can travel outwards at hundreds of kilometres an hour.

We are compensated though, when these particles collide into our protective atmosphere at the magnetic polar regions, giving us the spectacular light display we know as the aurora. The different colours we see relate to the various gases being ionised, such as oxygen (green), nitrogen (red) and neon (blue). Hey — it's nature's free laser light show!

Picture 21. The Aurora or Northern Lights.

It was once believed that the Sun shone by burning coal as its fuel source. However, this type of energy could only sustain it for a very short period.

Gravitational collapse seemed a possible answer which could generate enough heat, but even this idea failed to answer the question, how does the Sun keep shining steadily for such a long period?

Gravitational collapse could only last perhaps a few hundred million years at best. Knowing the age of the Earth to be much older, (from geological evidence and Darwin's Theory of Evolution) the Sun must be using some other means.

What could possibly be keeping the Sun shining with such regularity for literally billions of years?

The answer was finally revealed by unlocking the power deep inside the atom.

Atoms

The word atom appears to be a modern term, yet it was first coined as far back as 460 BC by the Greek philosopher Democritus.

Picture 22. The basic structure of an atom.

The idea that everything was made from small particles (atoms) was ignored for over 2,000 years. People preferred to hold onto the belief that the world was simply composed of Earth, Fire, Air and Water. Later on it was suggested that there must be an "atom" for each separate thing, one for plants, one for water, one for wood and even one for human flesh!

The first serious attempt to explore the structure of atoms was in the 1800s, and even then it took over 100 years before we truly began to understand it more fully. The more the atom was explored the more particles it seemed to reveal, leading scientists to refer to it at one time as a "particle zoo".

Discovering what makes up an atom is no easy task. They are so incredibly small, that you could fit 10 million of them along the edge of a postage stamp. Dealing on that minute scale is hard enough, yet only one millionth of a billionth is actually solid matter, the rest is virtually empty space. Various objects appear solid to us, yet it's really the repulsion of the electrons in the atoms that makes you feel things are hard or impenetrable.

If you could take away all the space out of the atoms, from all the people in the world (nearly 7 billion), you could fit everyone into the size of a sugar cube. Now how sweet would that be?

There are many different types of atoms. Those that are all the same kind make up the elements such as hydrogen and oxygen. Molecules are formed when more than one element combines, water (H**2**O) being one example.

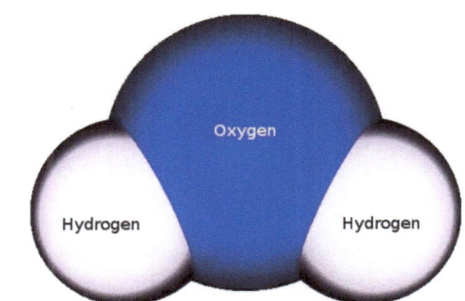

Picture 23. Hydrogen easily combines with oxygen to make a water molecule.

Molecules can also chemically bond, forming compounds. Table salt (sodium chloride) is a commonly recognisable compound. Although sodium is explosive on its own and chlorine is a poisonous gas; when united, we can end up safely eating them!

Everything you see around you, from the rocks in the earth to stars and planets are comprised of atoms. Whether you end up as a tree or a human simply depends on the type of atoms, and the molecular structure in which they are arranged.

It was once believed that atoms were the smallest particles that existed, but today, we know that even the atom can be broken down into much smaller divisions.

Apart from a few exceptions, atoms cannot be destroyed but instead are simply redistributed. When a tree burns down, the atoms don't disappear but become part of the air, just as the atoms of a dying plant rearrange themselves to become part of the soil.

Atoms, as mentioned, get recycled again throughout the universe, so you could have actually inherited some which were once part of an extinct dinosaur — makes me feel old, suddenly!

Your body contains around a billion billion billion (1,000,000,000,000,000,000,000,000,000) carbon atoms which were initially manufactured inside stars.

Stars

In 1967 an alien message was picked up from the depths of space. Kept secret from the public, it was classified as LGM or 'little green men'. The regularity of this signal was surely proof of other life forms out there.

The mystery was finally resolved, when it turned out to be a pulsar or very rapidly rotating neutron star. As the core spins, it gives off a beam of radio waves (similar to a lighthouse beam), and its signal (pointing towards Earth) was actually what was being picked up. Due to the regularity of its pulse, it was assumed to be a signal from some alien beings, as nothing like this had ever been detected before.

The lady who discovered them was not honoured, but instead, her bosses took the credit and were awarded the Nobel Prize — typical eh?

Picture 24. The comparison between a rural sky (top) and the view from near a city, affected by light pollution.

On a clear night, away from city lights, you might see around 2-3,000 stars, yet there are at least 200 billion stars in our Galaxy alone and countless other galaxies in the Universe. This means that there are probably more stars than there are grains of sand on all the beaches of the United Kingdom!

Stars are vast balls of hydrogen gas which are being converted into helium, just like our Sun, and in this process they give off energy, mainly in the form of heat and light. In the core of a star the extreme pressure and heat forces atoms together, and basically it's this nuclear energy that causes a star to shine, unlike planets which only reflect light from other stars.

Picture 25. The constellation Orion 'The Hunter' containing the red giant star Betelgeuse and the blue giant Rigel), along with the sword or pinkish nebula below the three stars of Orion's belt.

Stars vary in size, brightness, temperature and colour. One of the largest stars, Eta Carina, is over 2 million km across. To get some idea how big that is, imagine flying an aeroplane at 900km an hour over its surface. It would take you over 1,000 years just to go all the way round it.

The Pistol star is another enormous star which shines with a power equal to 10 billion Suns. However, it will only last for a few thousand years, compared to billions of years like our Sun.

This is because the more massive a star, the quicker it uses up its fuel source. When this happens, a star can expand enormously as its outer layers swell up and cool, turning red in the process. We aptly named these stars red giants.

Betelgeuse in the Orion constellation is one such star. If you placed it where our Sun is situated, its surface would probably extend out to the orbit of Mars.

It's expected to become a supernova (exploding star) at some point in the future, as its nuclear fuel is used up. That could even occur tomorrow, and when it does, it will radiate so much energy it will be bright enough to be seen in daylight, even although it's 427 light years away.

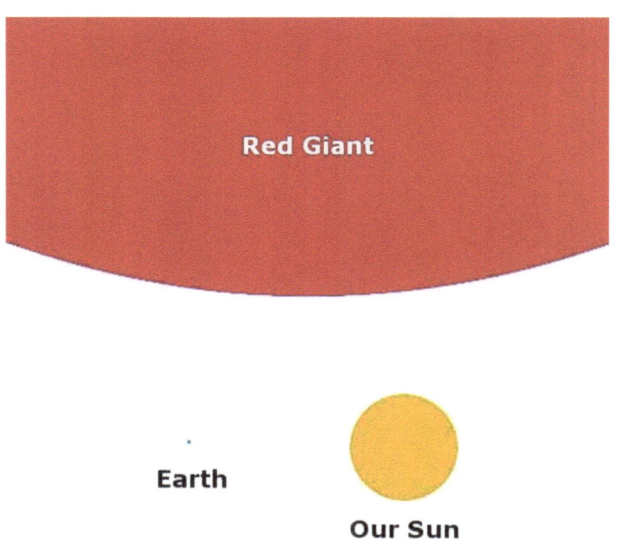

Picture 26. In comparison to some red giant stars, our Earth would appear almost invisible.

We often think of stars as being points of white light, but look closer and you will soon detect stars of different colours. Just as a hot flame appears blue, or a smouldering camp fire glows red, the colour of a star is a clue to its temperature.

Picture 27. Not all stars are white, as we often imagine.

The hottest stars are blue supergiants or hypergiants with temperatures around 50,000°C. They can be anywhere between 10 and 50 times the size of our Sun, and are thousands of times brighter.

Picture 28. This photo shows a star-like object (bottom left) but it's really a powerful supernova shining as bright as the central bulge of the nearby galaxy, which probably contains many millions of stars.

So how do astronomers know what a star is made of when they can't exactly go there to test it out? — The answer lies in examining the light from the star. Light is absorbed by the various chemicals the star contains, producing dark bands that can be read like a fingerprint or bar-code.

Picture 29. A spectrograph from the light of a star showing the various dark bands (Fraunhofer lines) indicating its chemical composition.

The atoms in stars are composed mainly of empty space, so gravity can squeeze them smaller until only the nucleus is left. This occurs in many stars at the end of their lives. There is always a balance between the outward pressure of the star and gravity pushing down. As the star's fuel source is used up, this equilibrium is altered and gravity can win, crushing a star down to many times its original size.

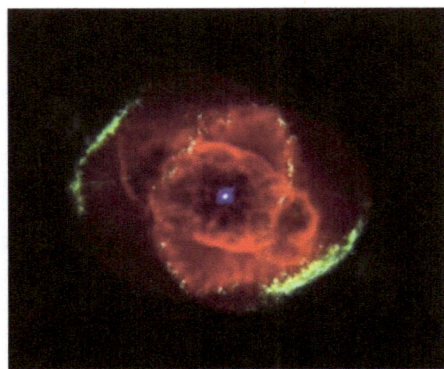

Picture 30. The catseye nebula. The remnant of a supernova explosion showing the remains of the neutron star at the centre.

The most extreme temperatures recorded have been found in neutron stars. These are the remnant cores of stars which have exploded in a violent supernova. As the stars core gets rapidly crushed by gravity, it can heat up to a staggering 1 million degrees Celsius.

They have a mass of about twice the size of the Sun, but are compressed down to a space of approximately 15km in diameter. This means that they are so dense that a small bucketful of material from a neutron star would weigh more than all the cars in the entire world!

Picture 31. The crab nebula in Taurus is an interstellar cloud created from a supernova remnant.

Stars larger than 4-5 times the mass of the Sun usually end their lives as black holes. These are collapsed stars in which gravity is so strong that even light cannot escape. It is now accepted that there is a black hole at the centre of every galaxy, including our own Milky Way. They are powerful enough to tear stars apart, cause galaxies to alter course and even affect space and time itself.

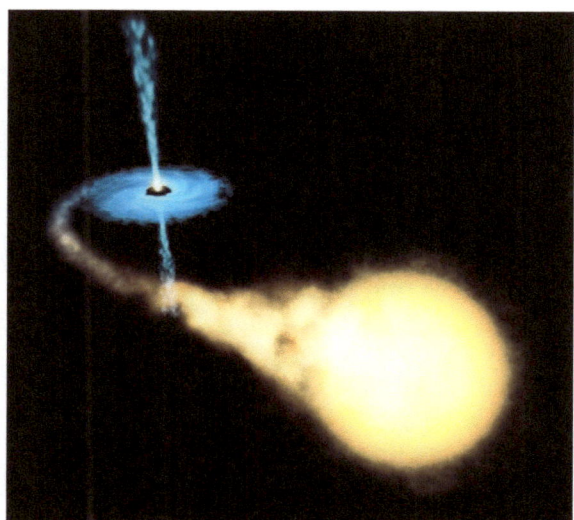

Picture 32. A black hole stripping material from a nearby star due to its immense gravity.

The largest black hole discovered to date is of apocalyptic proportions, being 18 billion times the size of the Sun.

These gigantic monsters seem to be terribly destructive, yet some astronomers believe that they now play an essential role in the formation of stars, galaxies and even life itself.

Stars to me are living things. Just like us, they are born, live and die. As stars like the explosive supernovas end their lives, they seed the universe with all the elements you will find here on Earth, from the aluminium in your drinks can, the gold in a bracelet, to the calcium in your bones or even the sodium in your blood.

Yes, folks, it's true — they were all manufactured in our sky's dazzling disco!

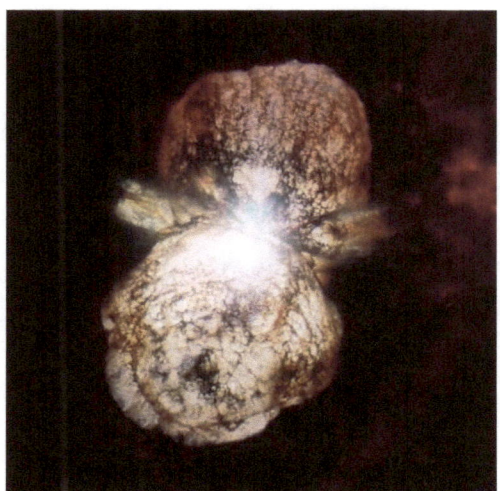

Picture 33. A star which has used up all its energy resources can ultimately explode, throwing out elements which will create the next generation of stars and planets.

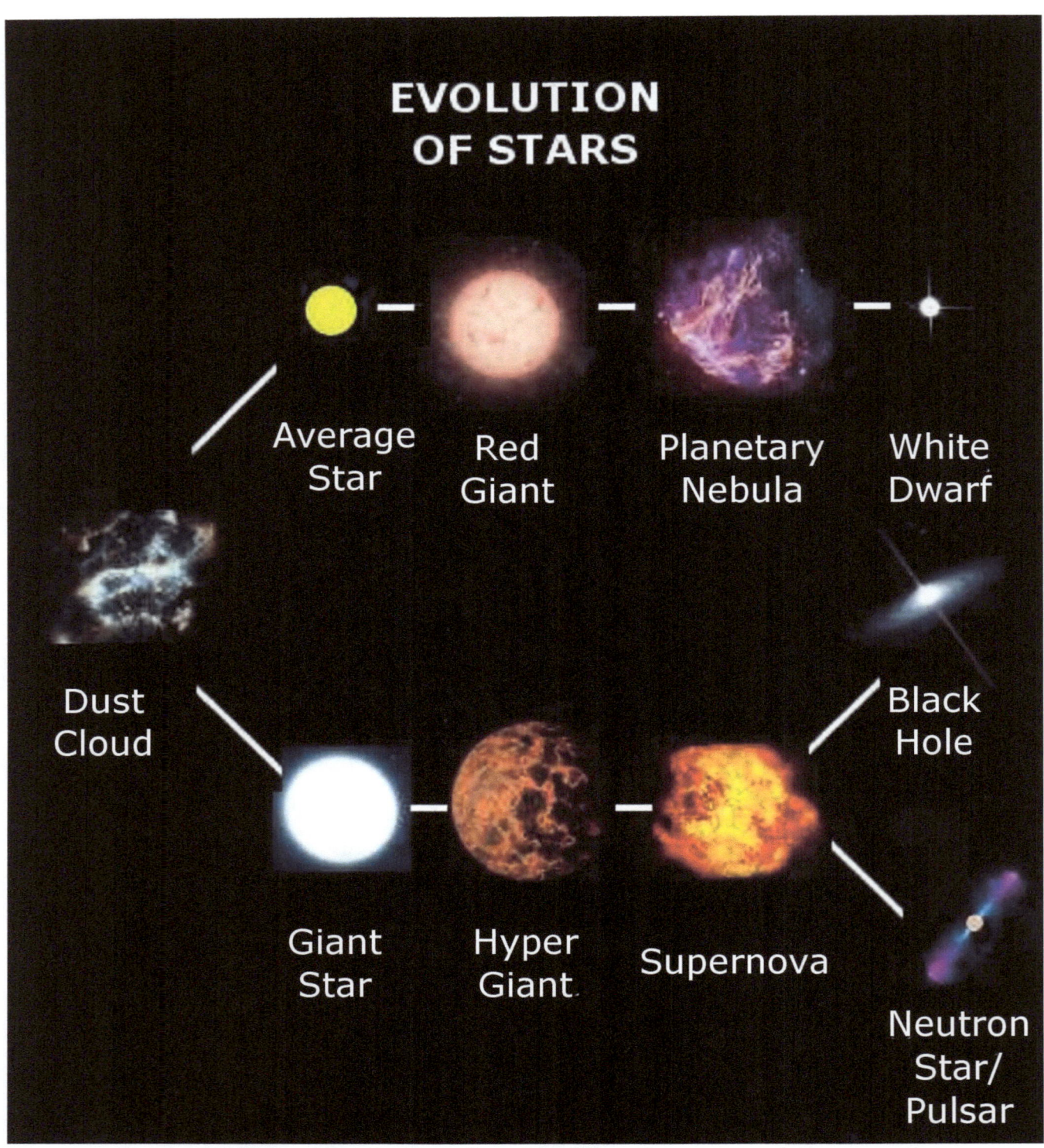

Picture 34. Stars about the size of our Sun follow the upper evolutionary path, and larger ones, the lower.

Have you ever wondered how astronomers know the distances to the stars? It's very complicated, but they use two different methods, one for nearby stars and another technique for those lying further away.

For stars up to a few hundred years away, a method called parallax is used. If you hold your finger out in front of you, then close one eye, and alternate that with the other eye, your finger will seem to move with respect to the background. Astronomers use this same "shift effect", and by using a bit of triangulation (measurement involving angles) it can give them a good estimation of a star's distance.

More distant stars are calculated by analysing the light from them, their brightness, colour, the various motions they display and even the regularity of the light given off from certain types of stars. It also involves trigonometry, geometry and the inverse square law, to give you some idea of the complexities entailed. They even employ satellites to help in the calculations!

Large distances of millions of light years can only ever be estimates, but are close enough to be fairly accurate.

So you see, they didn't use a tape measure after all!

Well, I can't end this chapter without mentioning a star that should interest most. Affectionately known as 'Lucy', this star is a white dwarf, or the core of a dead star that is now cooling down.

What's special about Lucy, is that it's mainly composed of crystallised carbon. Doesn't sound that enticing, but crystallised carbon is another term for a diamond, and therefore this star is valued at a billion trillion trillion carats! So you see, it probably was worth mentioning after all.

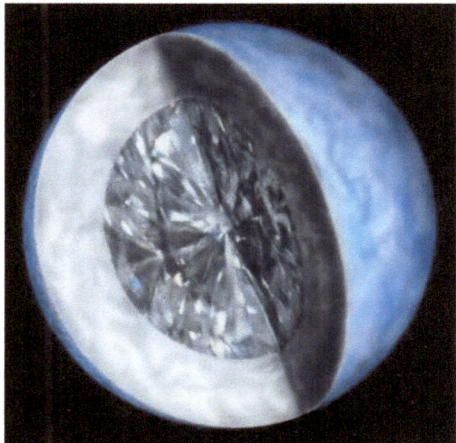

Picture 35. Artist's impression of the interior of "Lucy".

Space

Picture 36.

The word space has many meanings, but I tend to think of it as being anything outwith our atmosphere. It also reminds me of the great achievement of conquering space travel, which culminated in landing humans on another celestial body. That venture to the Moon really demonstrated the incredible ingenuity of mankind and has benefited us immensely, by advancing our scientific knowledge and giving us many new technologies.

If we hadn't gone into space, there wouldn't have been any smoke detectors, freeze dried foods, non-stick frying pans or satellite TV dishes. There have been over 30,000 by-products developed from space research which have helped to make our lives more comfortable. Just consider the help GPS systems have been to sailors navigating the vast oceans, anti-icing systems to pilots on aircraft or even running shoes for us, whose "springy" insoles were based on Moon boot technology.

That's definitely "one giant leap for mankind" Neil Armstrong didn't foresee!

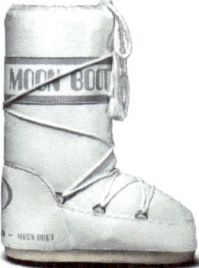
Picture 37. Moon boot available at www.moon-boot.com.

The novel Whiskey Galore, is a story about a ship whose cargo of liquor was washed ashore after a storm, giving the locals an unexpected alcoholic bounty to enjoy.

However there is another place that I know about, where you can find enough free booze to supply a whole nation for millions of years to come.

Sadly, this cosmic distillery lies within an interstellar cloud in the void of space. It's a cloud containing ethyl alcohol, or the same ingredient we use to make beer.

Be warned though, should you ever chance to drink it there are other nasty chemicals mixed in (hydrogen cyanide, ammonia and carbon monoxide) which are guaranteed to give you one very severe hangover!

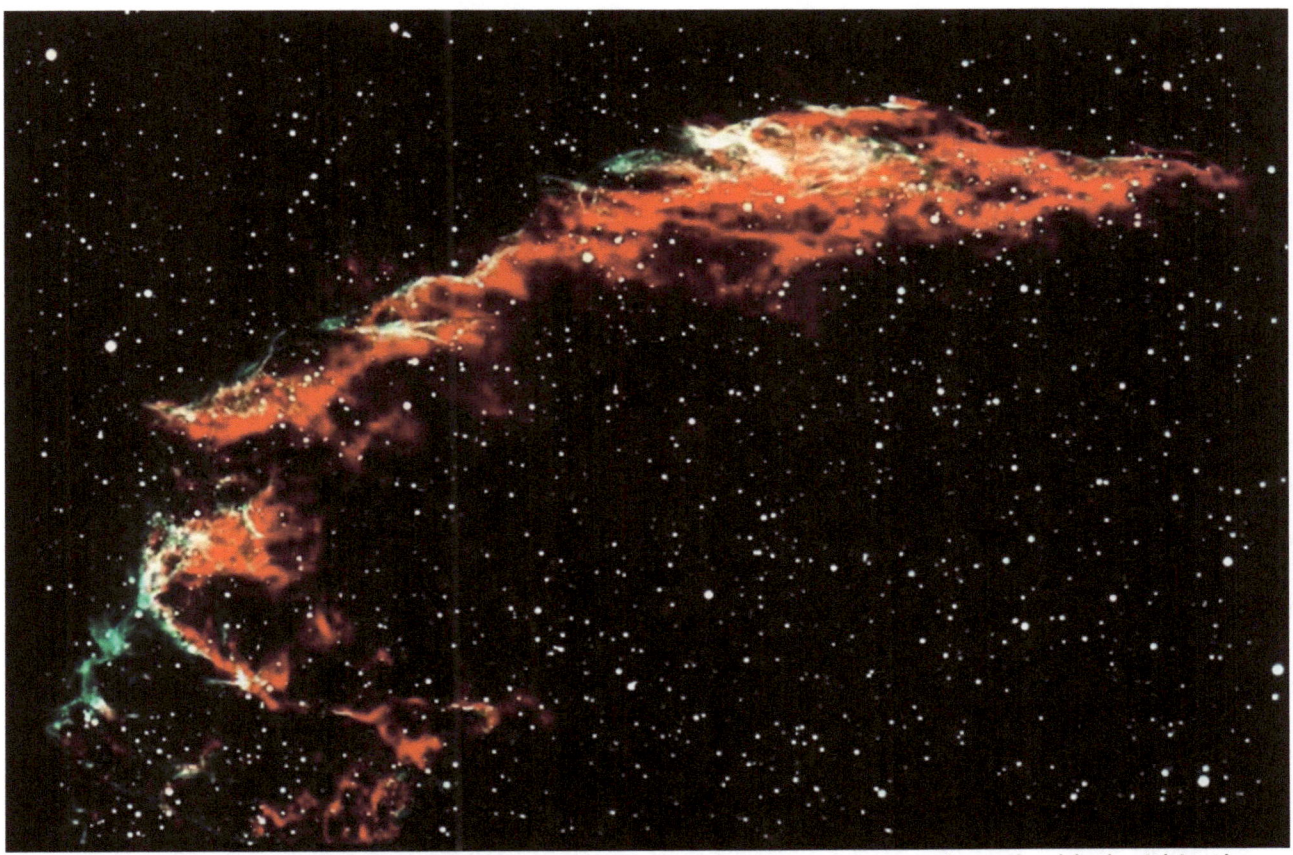

Picture 38. Some interstellar clouds have molecules not found on Earth embedded within them.

Imagine looking out of the window and seeing the Sun setting in the East and 90 minutes later watch it rising in the West. Sounds impossible, yet this is exactly what some astronauts have experienced as they orbited the Earth anti-clockwise.

I wonder how they first advertised for an astronaut's job — perhaps they enticed them by offering "good pay, free uniform, own cabin and free paid trip around the world." Hmm?

The Russians were the first to put a man into orbit, but I prefer to consider the American Joe Kittinger as being my first hero in space. In 1960 he ascended in a balloon to 31,300 meters, or 30 kilometres into the stratosphere.

What he did next though was truly amazing — he jumped! No he wasn't committing suicide, doing it as a stunt or because he was crazy.

Picture 39. Joe Kittinger's unique space jump.

This was a serious attempt to understand the problems of being in a space environment. His fall accelerated him close to the speed of sound, breaking 4 world records in the process, including the highest sky dive.

At that height, without protection, the body can literally "boil" due to the low pressure, yet he still went ahead with a ripped glove that had caused his hand to swell up tremendously.

Although his record might one day be broken*, I have to admire such bravery, because he didn't just do it once. He did it three times!

Admiration also goes to those early pioneer astronauts. I can't imagine too many people wishing to sit atop a big tube of highly explosive gas, knowing that it's never been tried before, and that once lit, there's no stopping it. There seems little comfort in being told "don't worry, we're all behind you" — yeah they would be — literally!

Back then, any voyage into space was full of many unknown dangers, along with a multitude of risks, which became particularly apparent with the later Apollo 13 mission. Only some ingenious ideas and a combined team effort helped these astronauts to return home safely.

---

* In fact this record has now been broken by Felix Baumgartner, who achieved a height of approx. 39km on 12 October 2012.

*Picture 40. Saturn V — the biggest and heaviest rocket ever used.
It was as big as St Paul's cathedral in London or 111m tall, was 10m wide without the fins and weighed around 3,000 metric tons.*

Space is an alien environment and every problem had to be thoroughly considered before putting human lives at risk. Fortunately, they at least managed to solve the issue of human waste disposal — phew!

You would need to travel at 40,000km an hour, or over 11km a second, to escape Earth's gravity. A Saturn rocket (one used in the Moon landings) burns 300 tons of fuel a second just to achieve that goal.

That's like emptying an average swimming pool every 15 seconds (think of the cost if that was your own vehicle), so another type had to be devised.

The reusable Space Shuttle seemed to be a possible alternative, but even its days were numbered due to the sheer expense. It was the most complex machine ever built by humans, yet its computer system ran on just one megabyte of RAM. That's only 0.005 percent of the power of an X-Box 360. Unlike normal aircraft, the Shuttle doesn't have the luxury of landing with its engines running, which means it doesn't have the opportunity of making a second approach if something goes wrong — Jeez, that is scary!

Picture 41. I know they said "outdoor work" but isn't this going a bit too far!

Picture 42. Silhouette of space shuttle Atlantis passing in front of the Sun.

We tend to think of the space around us as being empty. However, even if you removed all the air, dust and atoms it wouldn't be totally vacant. The reason is that minute particles are always coming into existence everywhere.

Space is not simply composed of nothing, but as you will soon discover, it can bend twist or even stretch.

Strange as this might seem, space is the invisible, yet flexible fabric of our universe and is not static as Sir Isaac Newton believed.

It's hard to visualise something you can't see as being able to bend or twist, but this has been shown to be true. Objects with mass, such as stars or planets can bend or warp the space around them, just as they can twist or drag space along with them as they revolve.

We think of space and time as being two different things. However, Einstein proved that space and time are interlinked - called spacetime.

This isn't so strange if you think about it.

Say you want to meet a friend at a coffee shop. Obviously you would need to arrange to meet them at a particular place and at a set time. Without the connection of a place (in space) and a unit of time involved, your plan would fail and the coffee go rather cold!

In a strong gravitational field, or if moving very fast, space can become compressed or stretched and time can slow down.

To illustrate this idea, picture shining a torch at a device for measuring the speed of light (let's call it a "speed camera"). The "camera" will obviously just record the speed of light.

Now imagine traveling towards the "camera" in a car at 100 kilometres an hour and then shining the torch. You would expect the device to register the speed of light plus the speed of your car, yet this doesn't occur and the "camera" always just displays the speed of light.

So what's going on, that the answer is always the same regardless of your motion? Something has to change, and believe it or not, but it's space compressing (length contraction) and time slowing (time dilation) that ensures that the speed of light remains constant.

Picture 43. Length is contracted in the direction of motion when traveling at very fast speeds and time is slowed down.

This contortion of space isn't something we are familiar with in our everyday experience, but it does happen when moving extremely quickly. Additionally, in the region of extreme gravity (a black hole), time can slow down so much it can stop altogether.

Einstein understood all of this stuff almost a hundred years ago, yet most of us struggle getting to grips with such abstract concepts. Anyway TIME to move on, as this is becoming more mind-bending than SPACE-bending!

Time

Picture 44. A beautiful Mayan calendar — around 1,000AD.

Time has always played a crucial role in our civilisation, yet the concept of time appears to be a human manifestation and not an intrinsic part of the universe at large.

Is time just a convenient tool we invented, a universal law, or is it simply an illusion? To ask the question — What time is it? — will likely invoke a multitude of answers, depending on who you request it from. It might be 7.20 pm on a Sunday in London, 14:20 to an American in New York, but it's also 04.20 am on Monday to an Australian in Sydney.

So time is not something as absolute and rigid, as we often perceive it to be. As previously explained, in the extreme gravity of a black hole it can slow down or even stop.

Time is something we live with and accept as a normal part of our lives. Time seems to regulate us, yet it's not something in universal terms that's always regular, as it can change. Is it just something we humans can't do without in our lives?

Picture 45. Time isn't always uniform.

Time is a strange thing that seems to control us, yet it isn't actually real if you think about it. It's obviously useful in our everyday lives but it makes you wonder why we attach such importance to something we just appear to have made up.

The first evidence of timekeeping stems back to 20,000 years ago when people carved markings on sticks and bones to record successive intervals between the new moon.

To talk about history itself though, implies some measure of the passage of time.

People also needed to know when was the best time to harvest their crops, take note of important events or prepare for impending weather. This led to the introduction of calendars which we are familiar with today.

Picture 46. A sundial was an early attempt to measure time.

Over the years, many devices have been invented to measure time, including water clocks, sundials, marine chronometers and the humble hourglass.

Picture 47.

Today we have atomic clocks that are accurate to within a thousand millionth of a second and only require minimal adjustment over millions of years.

Using these instruments, it was found that clocks aboard GPS satellites were running 45 microseconds slower than on Earth. This was further proof of Einstein's "Special Theory of Relativity" that time goes slower the faster you go or in a gravitational field. Strange as this might seem, it really does happen.

Taking this idea further, it means that if you set off on a 15 year journey from Earth, traveling near the speed of light, your clock would appear to run normally. However, on your return, your friends that were left behind would have aged considerably compared to you. Now you know the real secret of eternal youth!

Picture 48. The lamp in Pisa Cathedral which inspired Galileo.

While observing a suspended lamp swinging from the ceiling of the Pisa Cathedral, Galileo noted that it moved with great regularity, irrespective of the width or arc of its swing. He realised the potential of his discovery as a reliable method of timekeeping. Unfortunately someone else beat him to producing it commercially.

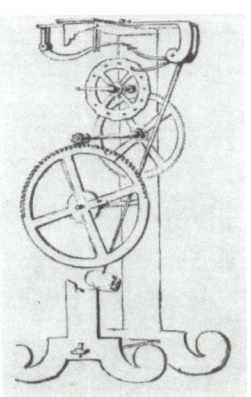

Picture 49. Galileo's original design for a clock.

It was only when aged 77 and totally blind that he finally related his design to a friend for his pendulum clock to be drawn up. Sadly he died before it was fully completed. What's remarkable though, is how he timed the swinging lamp, when there was nothing around then to measure such a small movement.

How did he manage it? Apparently, he used his biological clock — his pulse!

Galaxies

Picture 50. Our galaxy would look similar to this if viewed from above.

Most people ive in cities with bright lights which give them a very restricted view of the night sky. What might they see if all the lights were switched off?

Picture 51. The Milky Way, which contains billions of stars.

Stretching across the heavens is an unmistakable river of light we call the Milky Way. It's our edge-on view of the galaxy in which we live. Around 100 years ago, it was assumed to be the only one in the universe.

It was in the 1920s, that the astronomer Edwin Hubble discovered the first "island universe" outwith our own. Within a short period, billions of other galaxies had been located and we soon realised that we are not unique and alone.

Picture 52. No — it's not a dustbin floating in space but the Hubble telescope, which has managed to photograph galaxies lying an incredible 13 billion light years away.

How did the astronomers not realise there were all these other galaxies out there? Surely they must have been "like babies screaming out loud in an old folks' home", but sadly they just didn't have powerful enough telescopes to see them.

Today, we know that our galaxy is simply one of billions that pervade throughout the cosmos. Our galaxy is 100,000 light years across or 200 million solar systems wide, yet even that shrinks in comparison to one which is 6 million light years across or 60 times larger than our own.

Picture 53. Two galaxies, which to me resemble a romantic rose.

Galaxies come in a variety of shapes, some are elliptical, lenticular or irregular. Then there are galaxies which emit energy in the form of X-rays, gamma rays or even radio waves.

Picture 54. How our galaxy might look edge-on from the outside.

One particular type called quasars, radiates energy equivalent to the power of a trillion Suns. It's believed that this energy comes from the "in falling" material of a black hole.

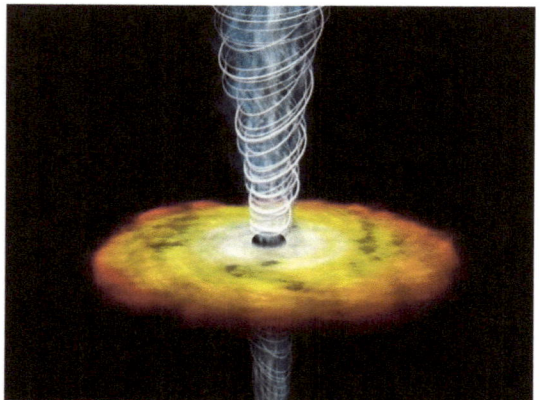
Picture 55. Impression of a quasar. It's believed that material falling into a black hole creates the intense light we can see, even although most quasars lie many millions of light years away.

Our Milky Way has a black hole which is 4 billion times the mass of our Sun. The largest black hole discovered to date is the size of 20 billion Suns or 150 billion kilometres across.

These giant monsters can gobble up stars or anything else that ventures too close. However, many astronomers now believe that they are not entirely destructive, but play an essential role in the formation of galaxies.

Another discovery by Edwin Hubble was that galaxies are moving away from each other very rapidly. In other words, the universe is expanding, or more correctly, the space between the galaxies is being pushed apart.

It appears, this is not too dissimilar as to how cosmologists believe the universe itself was created. Known as the "Big Bang", there was a sudden inflation of space that culminated in all the matter we see around us today. What's causing all this expanse is still one of the mysteries of science, so for now, it's simply termed dark energy. Nobel prize anyone?

Picture 56. The creation of the universe occurred in less time than the blink of an eye.

Galaxies are normally formed around nebulae (interstellar clouds) or in areas containing clusters of dust and gas or sometimes even stars and planets. It's here that stars, gas and dust become bound together gravitationally.

Picture 57. Looking like an ant's head, this odd looking galaxy is called the antennae galaxy.

Picture 58. Known as the "Pillars of Creation", these interstellar clouds span many hundreds of light years across.

However this is not the whole picture, as there appears to be another force acting to bind galaxies together. Although little understood, this stuff called simply dark matter seems to be a sort of "glue" that's keeping these structures together. Another Nobel prize on offer it seems!

Andromeda is one of our nearest galaxies. Lying about 2 million light years away, it's the furthest away object that can still be seen with the naked eye. It is one of 64 galaxies that are part of our local group, as most tend not to live in isolation.

Picture 59. The Andromeda galaxy, which will collide with our own Milky Way in about 4 billion years.

The largest structure of galaxies is known as the Great Sloan Wall and encompasses an area 1.37 billion light years across.

If you could view all the galaxies from the outside, they would resemble brain cells connected by filaments or strands.

Picture 60. The Sombrero Galaxy.

Picture 61. Two galaxies on a collision course.

Quantum Mechanics

Picture 62. The famous equation, yet few truly understand its real meaning or significance.

Before you go running off or skipping this chapter because of the title, you might want to know about the "weirdness" that occurs at the quantum level (the realm of the very small) and consider the fact that without it, you wouldn't have the luxury of using mobile phones, playing computer games, eating vegetables or basking in the warm sunshine.

So what is all this weirdness I'm referring to and what's it got to do with electrical devices, plants or the Sun?

Imagine having the ability to walk through walls, be in multiple places at the same time, pop in and out of existence or only be somewhere if someone decides to look. This might sound like some science fiction story, yet this actually happens all the time at the sub-atomic scale. Don't worry if you find this baffling, as even Einstein found it hard to accept initially.

It was discovered that these tiny particles act in very strange ways. One of these oddities is that electrons don't have a definite edge or an exact position. This means it's possible that they could be this side of a barrier, inside of it or even over the other side of it.

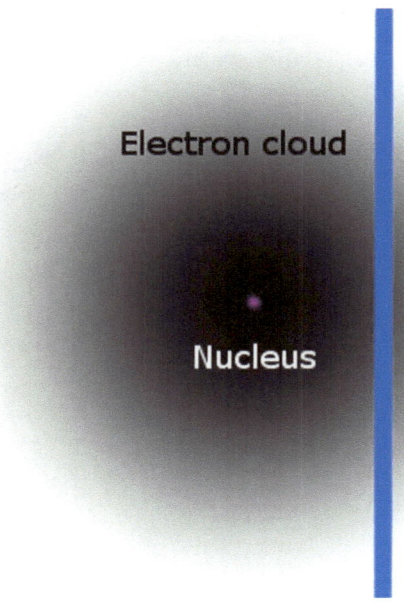

Picture 63. As an electron has this "fuzzy edge" to it, you can see how the electron has actually penetrated the barrier (shown in blue) and passed across to the other side.

Due to the wavy nature and positional uncertainty of an electron, if it can get close enough to a barrier, it has a good probability of getting all the way through to the other side.

This phenomenon, called quantum tunnelling, where particles can literally pass through obstructions unheeded, is an essential feature which allows electrical current to flow in transistors, and is the basis of all devices such as mobile phones and computers.

The same "tunnelling" mechanism is essential in helping to alter the structure of protons (part of the nucleus of the atom) in the conversion of hydrogen to helium in the Sun. Heat and pressure alone aren't enough to kick-start the Sun to shine.

Even stranger (if you haven't already fallen asleep) is that particles can be in more than one place at the same time (superposition) and can literally take every route possible, all at the same time.

That's equivalent to being in a cave system and being able to explore every passageway simultaneously to find the quickest way out.

This same extraordinary feat is what helps plants to be 99% efficient in process of photosynthesis. Without this "superposition", the light energy couldn't be transferred quickly enough to reach the part that makes the chemical energy. Plants have been using this method of finding the fastest route, millions of years before we even coined the term quantum mechanics. They are so efficient, they achieve this task in a billionth of a second. This certainly puts those people who claim to be good at multitasking to shame!

Another bit of quirkiness to digest is that particles can appear out of nowhere and disappear just as rapidly. This occurs constantly in the space around you, although they're too fast and too small for you to detect. This spontaneous creation from nothing is how scientists believe the universe itself was created. I'm aware that all this is counter-intuitive to your everyday experience but this is exactly how the world works at the micro scale.

Picture 64. Like fireworks, particles are formed from pure energy in the space around you, but quickly get annihilated on meeting their opposite anti-particle.

Before leaving this odd domain of science, it was found that particles can also behave like waves (duality) and that the simple act of trying to observe them changes their behaviour.

In other words, when you're not looking they are spread out everywhere and only have a defined position if you care to look.

Taking this a stage further, current research into quantum mechanics has now cast doubt on the true nature of reality itself. It implies that only by observing something does it have a definite shape or even length.

This now begs the question — are you actually reading this book or is it just a figment of your imagination?

Picture 65. Is the true nature of reality just an illusion?

Today we rely heavily on the fundamental principles of quantum mechanics and their remarkable properties.

We have already developed many technologies that benefit humans, such as mobile phones, computers, CAT scan machines and lasers, utilising the various quantum characteristics they possess.

In the future, computers could be programmed to do multiple tasks employing the peculiar properties of superposition.

It's amazing to think that we live in a strange quantum world and that our very existence is dependent on it.

We have a great deal to thank for the weirdness of quantum mechanics, yet we rarely give it a second thought!

Planets

Planets are believed to be the remnants of the gas and dust clouds of a newly formed star which have clumped together over time.

Not all planets orbit their parent star and some even have more than one.

Imagine living on a planet where you could watch three sunsets every evening, residing on one where ice is hot or being on one which has no days or years and where night is your constant companion. These strange worlds are just a few of the recently discovered planets outwith our solar system.

Picture 66. Rough size differences of the solar planets.

The Solar System was formed around 4.5 billion years ago from an interstellar cloud which collapsed under gravity, gradually spinning into a disk shape around the newly formed star we call the Sun.

Picture 67. A dark disk surrounding a newly formed star. This is how we believe our Solar System was formed.

The heavier rocky elements, such as iron remained close to the Sun while the lighter gas substances like hydrogen were thrown further out, forming the gas giants of Jupiter and Saturn, for example.

Mercury — is the closest planet to our Sun, yet regardless of its close proximity, it might still harbour ice on its surface. Its deeply cratered surface is a clue to its violent history (that's nothing criminal) which was caused by meteor impacts in the early bombardment era.

Picture 68. The well cratered surface of Mercury.

One impact crater left over is named the Caloris Basin. It has a diameter of around 1,500km and is surrounded by mountains reaching 2km high. That means the crater is larger than the length of the UK, and it's mountains are nearly 6 times higher than the Empire state building in New York.

There is little or no atmosphere on Mercury and its searing daytime temperatures of over 400C make it a very hostile environment for human habitation.

Although it's the nearest planet to the Sun, it's surprisingly not the hottest. That title goes to Venus, which lies 40 million kilometres further out.

Venus — The gas clouds surrounding Venus stop any heat escaping, which is why is known as a "greenhouse effect". The thick clouds of carbon dioxide around Venus mean that temperatures can soar to over 400 degrees Celsius in the daytime. — Wow! talk about global warming!

Picture 69. The greenhouse planet, Venus.

Until it was mapped by radar, very little was known about the surface of Venus. It turned out to have very inspiring mountains and is the closest in size to the Earth.

You wouldn't wish to go there though, with daytime temperatures hot enough to melt lead. Along with its thick clouds raining down sulphuric acid and pressures of over 90 times that on Earth, it certainly makes it a most unlikely holiday destination!

There's one advantage though about living on Venus — because its day lasts longer than its year, it means you can celebrate your birthday with a party that's guaranteed to last the whole year long.

Earth — The most precious stone worth treasuring is a lovely blue rock we call Earth. Unique among the Solar system, it is the only planet we know about that sustains life. If only we could ALL view our planet from space, perhaps we would understand its true fragility and come together more as a species, instead of ending up fighting wars and killing each other.

Picture 70. View of Earth from the International Space Station.

There is a very thin border in which we are able to survive here on Earth. You can go 20km horizontally and you will not perish. However if you travel the same distance either downwards or upwards without special protection, then you begin to appreciate that narrow range for life.

Picture 71. Earth's delicate "thin blue line".

We need to look after our delicate environment and ensure it remains that way for generations to come. I think the following words express that view clearly — if we turn and look towards the stars, we soon realise that it's a long way to the nearest watering hole.

Picture 72. The inset is an enlargement of the tiny white dot you can see lying between the rings of Saturn. The photograph was taken from the Cassini spacecraft. In case you are wondering, that little white dot is planet Earth! — The word "humble" springs to mind.

What's lesser known, is that Earth is the only planet to have water in all three states (liquid, gas and solid). It's one of the densest planets (what can I say?), yet it would also be the brightest in the Solar System, if viewed from space.

Earth is the only planet with plate tectonics, without which there would be a build up of carbon in our atmosphere (from the dead plants and animals in the sea) causing it to heat up too much.

Our planet is gradually slowing down due to the friction of the Moon, called "tidal locking". It means that in years to come, the days will be much longer than 24 hours. A few other unusual facts are:

- 1cm of rainwater falling on just one hectare of land would weigh over 100 tonnes.

- The Arctic (from the Greek *arktos*) was named after the constellation the Great Bear, not because of its Polar bears.

- The oldest rocks are found in the Highlands of Scotland and are around 4 billion years old (had to mention that, being Scottish).

- Lastly, there are at least a trillion trillion snowflakes that fall from the sky each winter. As each snowflake is truly unique, as well as beautiful and intricate, that's a mere 1,000,000,000,000,000,000,000,000 snowflakes to study. Hmm! — maybe not today though!

Picture 73. Snowflake crystal, taken with an electron microscope.

Question — Why is it hotter in Summer than in Winter, when we are actually closer to the Sun in the Winter months?

Picture 74.

Answer — The tilt of the Earth is facing more towards the Sun in the summertime, whereas in winter it faces away, therefore the rays striking the Earth are less intense.

Strange bright lights seemed to appear from nowhere, misty green shapes danced around wildly and Earth was plunged into darkness — no this is not some horror story but just a description of some natural events you can see here on Earth — a meteor shower, the aurora, and an eclipse of the Sun.

It's not surprising why our ancestors, seeing these things for the first time, believed them to be omens or the wrath of the Gods.

Picture 75. Tycho Brahe.

We owe a debt to the many famous scientists who helped further our understanding of the world, Einstein, Newton, Faraday, Alfred Nobel and Galileo for example, but one name I remember, more for his oddity, was Tycho Brahe.

He is noted for his brilliant observations of the planets, devised many precision astronomical instruments and was the first person to explain a supernova.

As a child, he was kidnapped from his parents by an uncle who then brought him up.

At university, he lost part of his nose in a fight with a fellow student about who was better at maths. He had to wear a prosthetic copper nose after that incident which even made him look rather weird.

Due to his interest in astrology, he employed a dwarf (as they were then called) named Jepp — who he believed was clairvoyant and who oddly used to sit under his table at dinnertimes. Jepp was also used by him to act as a court jester to entertain his guests.

Brahe also kept a pet elk, which accidentally got drunk on beer one evening and died falling down the stairs.

After Tycho himself died, he was found to have high concentrations of mercury poisoning in his system, apparently from swallowing some during previous alchemy experiments. Some people even speculated that he was deliberately poisoned.

Tycho Brahe wasn't a very popular man, firstly because he married a commoner and the eight children he had by her, were viewed as being illegitimate.

Also he drank heavily and was very abusive to his tenants, to the point of throwing them in chains if they didn't agree with him!

Although he might not have been looked upon as "Mr Nice Guy", I can't help but admire such an eccentric or "mad professor", who I'm sure deep down was a borderline genius!

It takes the Earth over 200 million years to circle the centre of our Milky Way (traveling at around 800km an hour). This is called a "galactic year". Earth has done over 20 revolutions like this since it was first formed. However, humans have existed for less than 1,000th of a galactic year.

This means that if the length of your arm to your fingertips was the span of the whole of Earth's history, human evolution would fit into the size of your fingernail. This brief timespan of human history suggests that we are perhaps not as advanced as we sometimes portray ourselves to be.

The first person to measure the circumference of the Earth was the Greek Eratosthenes in 276 BC. Amazingly he did it using just the Sun's shadow and some simple maths.

He knew that the Sun's rays were directly overhead at 90° (on the summer solstice) in a nearby town called Syene. The shadow cast by the sun where he lived (Alexandria) was 83°. Finding the difference in the degrees - seven (83° from 90°), he then simply divided that number into 360° or roughly $1/50^{th}$ of a circle and then multiplied the distance (approx 800km) between the towns to give him the answer.

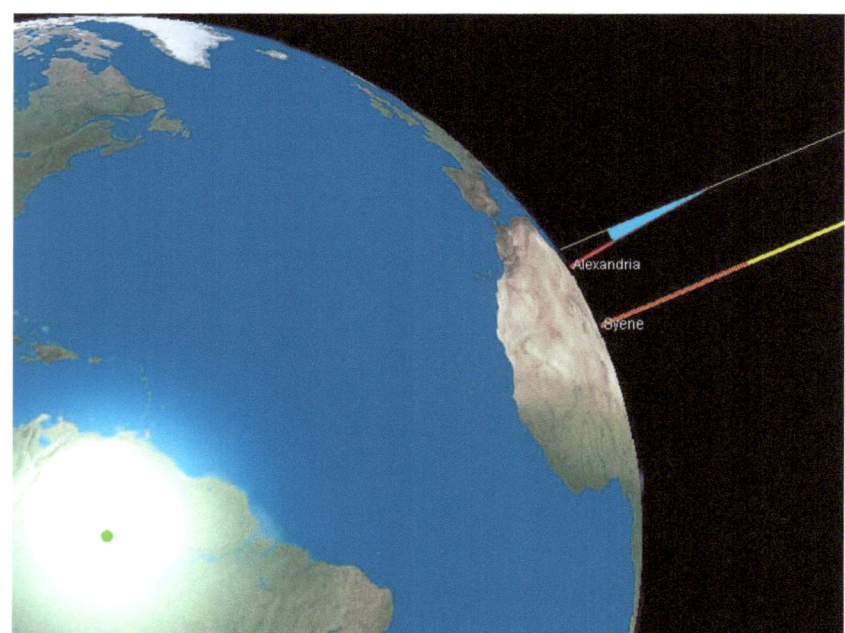

Picture 76. Eratosthenes' measurement of the size of the Earth.

His estimate was incredibly accurate and he was only about 400km out from modern measurements. That's pretty good for someone who lived nearly 300 years before Christ was born.

You can try doing this experiment yourself using a couple of sticks stuck in the ground at the right time of year. However, if that fails, then you could try rubbing the two sticks together to make a nice fire instead.

At this moment in time, you are traveling at over 1,000km an hour as the Earth rotates daily on its axis. You are also moving at nearly 100,000 kilometres an hour as the Earth revolves around the Sun. Furthermore, you are speeding along at 220km a second as the Earth makes its journey around the galaxy and even faster (600km a second) as the Milky Way orbits our local group of galaxies.

Don't you find it strange then, how some people still manage to turn up late!

Moon — Although not technically a planet of the solar system, being Earth's satellite, I decided to include it here.

Picture 77. Our natural satellite, the Moon.

Rock samples collected by the astronauts of the Apollo missions have led scientists to conclude that the Moon was originally formed from a violent collision with Earth by an asteroid or planet in our early history.

A geologist was on a visit to the Johnston Space Centre in Houston to study the rock samples retrieved from the Moon. While examining some of them, he asked the technician how old the rocks were estimated to be.

"They are 4.5 billion years, 2 months, 3 weeks and 1 day old" came the reply.

"How can you be so incredibly accurate in your estimation" the geologist enquired?

"Well I was informed that they were 4.5 billion years old when I started working here and I've been here now for 2 months, 3 weeks and 1 day".

Bet you thought I was being serious when you first started reading that part.

The Moon on average is about 384,000km away but it was much closer in the past than it is today. Records from fossils show that days were much shorter back then, and tides were far larger. Nowadays, scientists using lasers can accurately determine the distance to the Moon to within 23cm. Recent measurements verify that the Moon is receding from us at about 3cm a year, which means that in the future, people will not have the pleasure of witnessing the wondrous spectacle of a lunar eclipse.

Picture 78. The Sun's corona as seen at an eclipse. The temperature of the corona is many times hotter than at the Sun's surface. A mystery that's still to be resolved.

If you can imagine a motorway stretching from Earth to the Moon, and driving a car at 100km an hour non stop, how long do you think it might take you to complete that journey? Provided you had enough fuel, it would still take you over 5 months to reach your destination.

If you cared to walk there at 5km an hour, without any rest, then a trip lasting around 9 years awaits you! — While I go off to test that theory, see if you can solve this tricky puzzle —

Mr X claims to come from another planet, where Afternoon comes before Morning, Tomorrow comes before Yesterday and Later comes before Now. — I told him, this is nothing unusual and that you can find the same thing here on Earth. Can you figure out why my answer is perfectly correct? Go to page 68 for the solution.

Mars — In 1938 in America, widespread panic ensued when the radio news broadcast informed people that a Martian invasion was already underway. However, it was simply Orson Welles narrating the novel "War of The Worlds".

It seemed to be such an authentic news broadcast at the time that it disrupted households, stopped several religious ceremonies, clogged communication services, caused traffic jams, not to mention the fear it must have struck into all the small kiddies! It's ironic, but Mars has two moons called fear and panic or Phobos and Demos as they are known.

It was in 1877 that surface features resembling "canals" led people to question the possibility of alien life forms existing there. However, it turned out these canals were merely just an optical illusion.

Picture 79. Mars — must be very old. It's rusting!

In 1901 Charles Pickering wanted to create a series of mirrors on Earth, so we could signal to our alien friends on Mars. At least some people are still convinced of the existence of life there. Currently the Mars rover Curiosity should finally settle any argument of a Martian civilisation.

Mars does have a surprising terrain, having the highest volcano and valley in the Solar System. Olympus Mons is 27km high, which is three times larger than Mount Everest, and is 550km wide.

It's also the highest mountain that we know about to date, and the crater Valles Marineris is the deepest and longest in the whole of the Solar System. It stretches 3,000km long and the chasm walls plummet down 8,500m. That's equivalent to nearly the width of the USA and about six times deeper than the Grand Canyon.

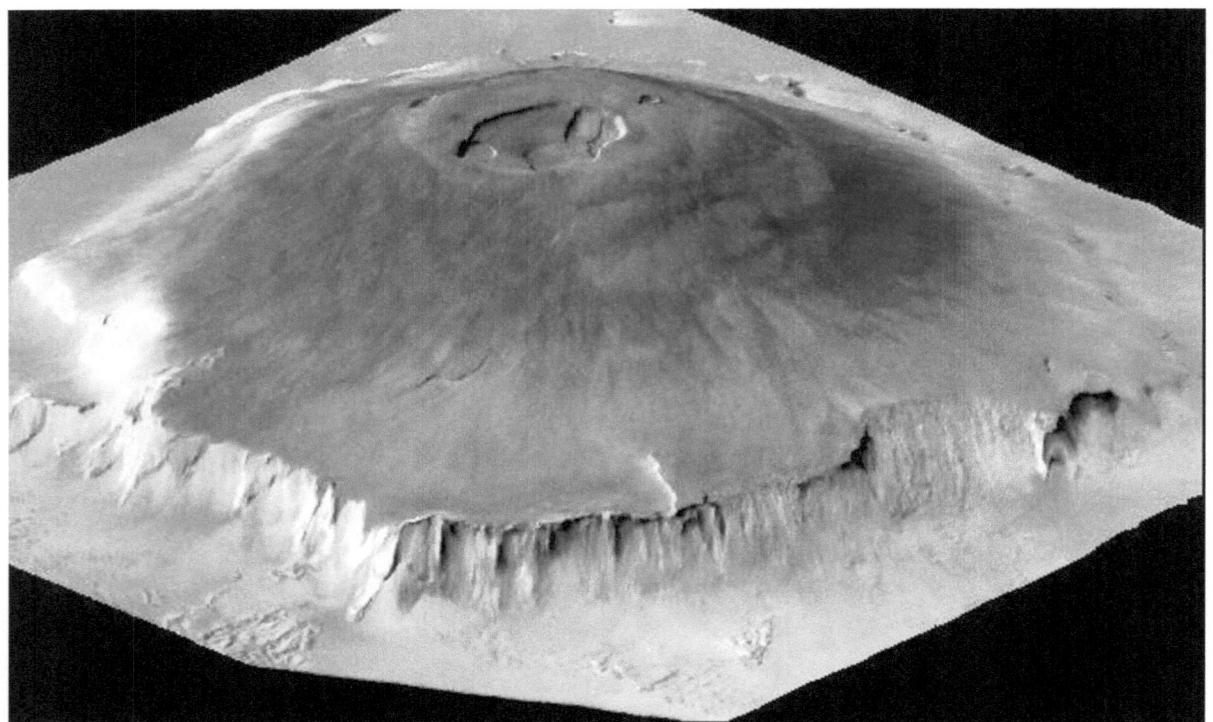
Picture 80. Olympus Mons — the highest volcano in the solar system.

Mars is known as the Red Planet due to its iron oxide content — or as we more commonly know it — Rust! Still, I love the names for some of the rocks there, such as Barnacle Bill, Cabbage Patch, Stimpy, Shark, Yogi and Pop Tart for example.

The Martian atmosphere is mainly CO_2 (carbon dioxide) and there are ice caps, deserts and even seasons on Mars.

What I like best though, is that you can jump three times higher on Mars than on Earth, due to its lesser gravity — now that sounds fun!

Jupiter — is the largest planet in our Solar system. It's 143,000km in diameter and big enough to fit all the planets of our Solar system inside it. It's the first of the "gas giants" out from Earth. Jupiter is 11 times the size of Earth.

More importantly, it has such a strong gravity that it's deemed to be the "hoover" of the Solar system. Anything coming too close will get pulled in, as the comet Shoemaker-Levi 9 found to its cost in 1994.

traveling at 216,000km an hour, the fragmented comet slammed into the Jovian atmosphere creating a fireball with temperatures of around 25,000 Celsius.

Furthermore, the shattered comet, totalling 21 pieces of about 1-2km wide caused seismic waves and created a dust storm the size of Earth. Still, this spectacle helped scientists to understand more about comet impacts and see just what damage could occur here, should one be heading our way.

Picture 81. Tell-tale impact signs (lower right) after the comet Shoemaker-Levi 9 plunged into Jupiter.

Jupiter has the fastest rotation of any of the other planets, going round in less than 10 hours. That's one very quick day for a large planet! It's this speed which causes the clouds on Jupiter to form bands, and is likely responsible for the Great Red Spot too, which we are familiar with. This is a vast storm system that's three times larger than Earth and has been raging for around 400 years.

In 1979, the spacecraft Voyager-1 was the first to take close up views of Jupiter. It's now traveling outwith our Solar system at a speed of 75,000km an hour and is traveling over three and a half times the distance to our Sun in a year.

Picture 82. Voyager, the most distant human-made object in Space.

It carries a phonograph record which has the sound of waves breaking, speeches from American presidents, greetings in 55 languages, the noise of thunder, a variety of music and even instructions on how to play it! I wonder what any future alien civilisations finding it will ever make of it all?

Saturn — is the beautiful ringed planet. When it was first seen in a poor low-powered telescope, it appeared to have ears. This was in fact the ring system but it's easy to see why it looked that way initially.

Picture 83. The beautiful ringed planet Saturn with three of its moons: Tethys, Dione and Rhea.

The ring system is about 180,000km wide or half the distance to the Moon from Earth. It's nine main rings are composed of millions of ice particles and rocks, some the size of a standard house.

Saturn lies 1.5 billion kilometres at its furthest point from Earth and is about 10 times larger. The word Saturday is derived from it and it's also named after the god of farming.

It's one of the least dense of the planets and if you could stick it in a large enough bathtub, it would actually float.

It is incredibly cold on the surface of Saturn, with temperatures of -212°C. The coldest temperature ever recorded on Earth was -89°C. However, Saturn's interior is surprisingly very hot at 11,500°C.

Winds are strong there and can rage at 1,800km an hour. That's about four times faster than anything recorded here on Earth. Saturn also has powerful lightning storms that can last for years and a magnetic field 600 times stronger than Earth.

You couldn't stand on Saturn's surface as it's mainly hydrogen and helium, and even if you could, you would be crushed by its fearsome gravity that's 100 times more powerful than on Earth.

If you could live there, at least one consolation, is that you can look forward to a summer that lasts for a full eight years.

Uranus — is most unusual, in that it spins on its side. So really it doesn't have a North or South as we know it, as North is in a westerly direction and South is in the East. So basically its North and South poles lie at the Equator — confused?

Picture 84. Uranus — the sideways planet.

Its sideways tilt also means that it spends 42 years bathed in sunlight and 42 years in darkness.

The tilt of Uranus was likely caused by a collision from another planet, asteroid or moon in its early history.

Uranus was the first planet to be discovered using a telescope back in 1781. It has 27 moons which are all named after literary characters from Shakespeare and Alexander Pope.

It's the 7th planet out from the Sun with a distance of 2.9 billion km, or 19 times further away than the Earth.

It has an atmosphere of hydrogen, helium, methane along with ammonia and even has liquid metal in it's core.

Unlike all the other planets, Uranus is named after a Greek God and not titled using the normal Roman classification.

It was always an odd name to me too and caused much embarrassment at school. It should be pronounced "Yoor Ae Nus" and not "urine nus!"

Neptune — Looking like a lovely blue Earth from space, it's anything but Earth-like. The winds on Neptune are the fastest we know about of any planet, with speeds of over 2,000km an hour or three times as fast as our strongest hurricanes. Its blue colour looks like an ocean but it's the methane atmosphere that's mainly responsible by scattering the light.

Picture 85. Neptune looks calm but it has incredibly powerful winds hidden from view.

It's appropriate that the ancients named it after the god of the sea and it's one of the most stunning planets due to its colour.

Neptune has 13 Moons, one of which spouts out nitrogen frost all over its surface. It has seasons, just like Earth, but because it takes 165 years to orbit the Sun, each season lasts a full 40 years.

Neptune is the coldest planet in our solar system, yet its moon Triton is even colder at -360ºC. I guess with it being 4.5 billion kilometres from the Sun we can't expect it to be hot exactly.

It might surprise you to know, that along with Uranus and Saturn, Neptune also has a ring system. Neptune was the first planet to be discovered using maths calculations to predict its orbit.

That's pretty smart, as the answer I gave in my maths test for the number of seconds in a year was twelve! — being the second of January, the second of February, the second of March …

Pluto — Aw poor wee Pluto — it was just too small to be considered as a planet any more — so they sacked it! Anyway, there are so many other bodies out there just beyond Pluto, in what's known as the Kuiper Belt and the more distant Oort Cloud, that contain much bigger chunks of rock.

Sorry Pluto! — but I can't include you in this section I'm afraid. What can you expect, when it took an 11-year-old girl to finally suggest the name Pluto, then I discover it stands for god of the underworld! — Jog on Pluto!

Picture 86. Pluto with its moon Charon.

Exoplanets — There are now around 1,000 exoplanets that have been discovered outwith our Solar system and there are new ones being added daily. Although no true Earth-like planets have been found, there have been many real surprises. Planets orbiting pulsars (high radiation stars) and ones that are so close to their parent star that the heat produces raindrops that are made of molten iron.

Picture 87. A raging volcano, which you could expect to find on any planet lying close to its parent star.

There are some exoplanets which are rotating around neutron stars or the cores of supernova remnants. I can only imagine that they were captured afterwards by the stars gravity, as no planet could ever survive the power of a supernova explosion so nearby.

There are planets orbiting several Suns and even some called "hot Jupiters" that have been found to spin in the opposite direction to their parent star. This discovery brings into question the true origin and formation of solar systems, which we had otherwise been led to believe.

Puffy planets are similar to hot Jupiters and get their name from their swollen atmospheres, caused by the heat of being in very close proximity to their main star.

There are even planets called "Super Earths" some of which could be covered entirely in water. If so, there might be no land surfaces to stand on but they'd be great places to go to if you're a sailor.

Picture 88. A "super earth" planet, possibly covered entirely in water.

Who knows what other worlds are still out there to be found. After all, there are billions of stars in our galaxy, and even more galaxies in the universe, all with the possibility of planets around their stars which could harbour life.

Mr X Answer: Stop cheating! Go back and think about it. But if you're really stuck, then go to page 72.

My favourite though has to be "Gypsy" planets or wandering planimos. They simply drift along without any parent star to orbit. I just like the idea of drifting around, not knowing where you might end up next. Much as that seems idyllic, these planets are very cold, dark and with no defined future. Sounds like someone I once met at the Job Centre!

Picture 89. Twin suns setting, as might be viewed from an alien planet.

Comets, Meteors and Asteroids

Picture 90. The Comet Hale-Bopp, which appeared in 1997.

Unlike Santa, we don't have Comets appearing yearly due to their very elliptical orbits. Some like Halley's Comet take 76 years to go around the Sun, but at least you can easily predict its next arrival.

Comets are "dirty snowballs" composed of ice, rock and basically dirt. Not all comets are simply made of ice though. One called "Amun" is mainly made of stainless steel. Its iron, cobalt and nickel content is valued at several billions of pounds sterling.

Some comets take as little as seven years to complete their orbit, while others can take thousands of years. As they approach the Sun, they "melt," shedding large dust trails that can stretch thousands of kilometres across the sky.

Most comets are believed to originate in areas lying outside our Solar System, which are then deflected from their orbits and captured by our Sun.

There are around 5,000 Comets we know about and their nucleus can range in size from 100m to 40km. As they approach the Sun, they can be traveling at over 1 million kilometres an hour.

In 1910, as a comet was due to approach Earth, a shrewd businessman managed to sell gas masks, umbrellas and anti comet sickness pills to those people who superstitiously feared the worst.

I wonder how he might have fared on "Dragons' Den".

Some comets break up or their "tails" can become future meteors. These fragments are often as small as bits of dust. As they hit our atmosphere they burn up, giving us those wonderful streaks of light across the sky we know as "shooting stars".

There are around 100 tons of space matter that fall onto the Earth every year which actually causes Earth to become heavier. Occasionally, a few bigger pieces manage to reach the ground and are named meteorites.

Picture 91. Asteroids come in many shapes and sizes.

Asteroids are generally the remnants of failed planets, which are pulled or nudged out of their orbit by gravity and captured by other stars or planets such as the Sun or Earth.

It's estimated that an asteroid collides with Earth every 50-100 million years and there are around 8,000 asteroids which are about 500 to 1,000kms in size that pass close to Earth.

Any asteroid plunging into Earth would cause incredible devastation. This is what scientists think happened 65 million years ago that wiped out the dinosaurs.

The power stored up from the lightning speed, plus the kinetic energy gained from the gravity of falling, has the power of a really huge nuclear bomb going off.

Picture 92. Shooting stars light up the sky but most are only about the size of a grain of sand. It's only their speed through the atmosphere that causes them to glow so brightly.

Intricate patterns you weave in the sky
Such wondrous vision passing me by
Touching my soul and refreshing my mind
Shedding your secrets in the dust left behind.

(Excerpt from the poem "Shooting Star" G.H. 1988)

A large asteroid hitting Earth could easily wipe us out, just as it did with the dinosaurs. No wonder scientists are interested in knowing more about them, with the view of avoiding any future devastating collisions with our planet.

Picture 93. What the dinosaurs didn't see coming.

An asteroid might be terribly destructive, but it now seems they could actually be responsible for bringing life to our planet. They contain traces of amino acids and organic compounds which may have seeded Earth and started off the initial process of life on Earth.

All these terms, comets, meteors and asteroids tend to be grouped together but what's been illustrated here are the main differences.

A NASA experiment, which involved exploding a comet to discover its interior, caused one Russian lady to become very upset. She went so far as to take the case to court. Her complaint was that it would interfere in her work as an astrologer and would affect her future horoscope predictions.

She claimed it violated her "life and spiritual values" and caused her "moral suffering". It seems she couldn't see the scientific value of such an experiment. Not surprisingly she lost her case, especially when it also turned out she was suing for 300 million US dollars!

Picture 94.

Mr X Final Answer: You will find them all alphabetically in a dictionary.

The Future

We can't predict with any certainty what the future ahead will be like, or even where new discoveries will lead us.

Picture 95.

However, it was only a few hundred years ago that we believed we resided at the centre of the Solar System and that our Milky Way was the only one in the universe.

Today, we are discovering worlds far beyond our own and are beginning to unravel the origin of creation itself. Advances in science and technology have propelled us from the Wright brothers' maiden flight to landing humans on the Moon.

As this rate of progress seems to be continuing, then I'm confident we will eventually find the technologies that will allow us to explore further out into the cosmos.

Discovering another habitable planet and going there is essential if our species are to survive.

In a few billion years, the Sun will swell up to become a red giant star, as its nuclear fuel is used up, and will eventually extinguish all life on Earth.

Meantime, we can look forward to new discoveries, enjoy the luxuries created by technology, and trust that we don't destroy ourselves before the Sun does.

Picture 96. This is Earth's ultimate fate, as the Sun will swell up to engulf it in about 5 billion years from now.

In the future, computers might use the properties of quantum mechanics to become many times faster, we will have robots doing some of the more menial tasks, trips to the doctor for diagnosis could likely become obsolete and new advances in medicine will help extend our lifespans considerably.

Artificial "iron man" suits to help the disabled to walk, invisible crash helmets, electricity from aluminium/water, eye-tracking systems to open computer pages and truly "long-life" batteries are some of the things already being tested.

Currently work on nanotechnology (manipulation of matter at the atomic scale) is one advancement that could drastically alter our way of life, through its future uses in many areas of society in the years ahead.

What I really want though, is a time machine, but I've been told by my friends, that I already have a watch!

Summary

I set out to write this book with the aim of making it a factual, yet also light-hearted book about the universe in general. I'm aware that it's a bit "patchy" in places and lacking in details. However, it was just meant to spark off the reader's interest and to hopefully inspire people to read more about the subjects covered.

I also wanted to highlight the importance of our link to the Universe and how new discoveries about it are influencing our everyday lives.

I talked about gravity and how it affects everyone and everything around us. I've dealt with light, and that without it we couldn't have known about our early origins.

Then there was space and time, the framework in which we live, along with the atoms in our body which we inherited from the stars.

I've shown how our lives are governed by the laws of physics and how comets could eventually destroy us, as will our nearest star — the Sun.

I hope it hasn't been all doom and gloom though, and trust you enjoyed this little foray through the pages.

We are connected with the cosmos in lots of ways, whether we care to think about it or not. I hope this book has demonstrated this and that you will be interested in learning more. As Albert Einstein famously said — "the important thing is to not stop questioning".

Picture 97. Einstein during a lecture in Vienna in 1921, aged 42.

Picture 98. Welcome back down to Earth. I hope you've enjoyed the journey.

If any readers have any questions or comments, I will be happy to answer them. If you write to me at Connection2Cosmos@yahoo.com then I will do my best to respond to you as quickly as possible.

Photo and Illustration Credits

Cover	NASA/CXC/JPL-Caltech/STScI	Picture 50	NASA Nick Risinger
Picture 1	Sanyamshri	Picture 51	Steve Jurvetson
Picture 2	NASA/ESA/STScI and Hubble	Picture 52	NASA
Picture 3	NASA	Picture 53	NASA/ESA and Hubble Heritage Team
Picture 4	NASA	Picture 54	NASA/ESA Hubble
Picture 5	NASA/ESO and M Livio	Picture 55	NASA/JPLcaltech
Picture 6	Oruben, NASA	Picture 56	Stephen van Vuuren from Saturn's Rings http://www.insaturnsrings.com/
Picture 7	Brian Blondel	Picture 57	NASA Brad Whitmore
Picture 8	LadyofHats	Picture 58	anonimo1
Picture 9	xcitefun.net	Picture 59	Adam Evans
Picture 10	NASA	Picture 60	NASA/ESA
Picture 11	Nick Strobel astronomynotes.com	Picture 61	NASA/ESA/Hubble Heritage Team
Picture 12	Created by User:Johnstone using a 3D CAD software package and an image of planet earth from NASA's Galileo spacecraft. http://en.wikipedia.org/wiki/File:Spacetime_curvature.png	Picture 62	Derek Jensen (Tysto)
Picture 13	Frann Leach	Picture 63	Frann Leach
Picture 14	Cam-Ann with mods by Frann Leach	Picture 64	Estormiz
Picture 15	Frann Leach	Picture 65	emijrp

Picture 16	Blue Marble by NASA with mods by person unknown	Picture 66	NASA/JPL
Picture 17	NASA Earth Observatory/NOAA NGDC	Picture 67	NASA
Picture 18	NASA/ESA/Soho-EIH Consortium	Picture 68	NASA/John Hopkins Uni/Washington Carnegie Institute
Picture 19	NASA/SDO	Picture 69	NASA
Picture 20	NASA/SOHO	Picture 70	NASA
Picture 21	Senior Airman Joshua Strang, USAF	Picture 71	NASA
Picture 22	Halfdan on wikipedia	Picture 72	NASA/JPL/Space Science Institute
Picture 23	Sakurambo with mods by Frann Leach	Picture 73	NASA/Ben-Zin
Picture 24	Jeremy Stanley	Picture 74	RH Castilhos
Picture 25	NASA/GSFC & MSFC Matthew Spinelli	Picture 75	unknown artist
Picture 26	Frann Leach	Picture 76	Lookang/Tom Patterson, www.shadedrelief.com
Picture 27	NASA	Picture 77	NASA/Sean Smith
Picture 28	High-Z Supernova Search Team/HST/NASA	Picture 78	NASA/Exploratorium
Picture 29	Saperaud	Picture 79	NASA
Picture 30	NASA, J. P. Harrington (U. Maryland) and K. J. Borkowski (NCSU)	Picture 80	NASA/JPLcaltech Mars Program
Picture 31	NASA/ESA J. Hester, Arizona State Uni	Picture 81	NASA/Hubble Space Telescope Comet Team
Picture 32	NASA/ESA	Picture 82	NASA
Picture 33	NASA/ESA	Picture 83	NASA
Picture 34	Frann Leach from pictures by NASA	Picture 84	NASA/Space Science Institute

Picture 35	Harvard-Smithsonian Center for Astrophysics	Picture 85	NASA/JPLcaltech
Picture 36	NASA/ESA	Picture 86	NASA/Pat Rawlings
Picture 37	www.moon-boot.com	Picture 87	Wolfgang Beyer
Picture 38	Astrogallery	Picture 88	NASA/JPLcaltech
Picture 39	USAF	Picture 89	NASA/JPLcaltech
Picture 40	Murali Dhanakoti	Picture 90	Andy Roberts
Picture 41	NASA	Picture 91	NASA/JHUAPL
Picture 42	NASA/Thiery Legault	Picture 92	Navicore
Picture 43	Frann Leach	Picture 93	istock
Picture 44	Truthanando	Picture 94	Toby Hudson
Picture 45	Frann Leach	Picture 95	Frann Leach
Picture 46	Liz West	Picture 96	FSGreggs
Picture 47	Martin Olsson	Picture 97	Ferdinand Schmutzer
Picture 48	Vicenzo Possenti	Picture 98	NASA STS-119 Flight Day 14 Gallery
Picture 49	Vincenzo Viviani		

www.ingramcontent.com/pod-product-compliance
Lightning Source LLC
Chambersburg PA
CBHW050853180526

45159CB00007B/2667